2020年

全国水利发展统计公报

2020 Statistic Bulletin
on China Water Activities

中华人民共和国水利部　编

Ministry of Water Resources, People's Republic of China

·北京·

图书在版编目（CIP）数据

2020年全国水利发展统计公报 = 2020 Statistic Bulletin on China Water Activities / 中华人民共和国水利部编. -- 北京：中国水利水电出版社，2021.9
ISBN 978-7-5170-9996-3

Ⅰ. ①2… Ⅱ. ①中… Ⅲ. ①水利建设－经济发展－中国－2020 Ⅳ. ①F426.9

中国版本图书馆CIP数据核字(2021)第197752号

书　名	2020年全国水利发展统计公报 2020 Statistic Bulletin on China Water Activities 2020 NIAN QUANGUO SHUILI FAZHAN TONGJI GONGBAO
作　者	中华人民共和国水利部　编 Ministry of Water Resources, People's Republic of China
出版发行	中国水利水电出版社 （北京市海淀区玉渊潭南路1号D座　100038） 网址：www.waterpub.com.cn E-mail：sales@waterpub.com.cn 电话：（010）68367658（营销中心）
经　售	北京科水图书销售中心（零售） 电话：（010）88383994、63202643、68545874 全国各地新华书店和相关出版物销售网点
排　版	中国水利水电出版社微机排版中心
印　刷	河北鑫彩博图印刷有限公司
规　格	210mm×297mm　16开本　3.75印张　61千字
版　次	2021年9月第1版　2021年9月第1次印刷
印　数	0001—1000册
定　价	39.00元

凡购买我社图书，如有缺页、倒页、脱页的，本社营销中心负责调换

版权所有·侵权必究

目 录

1 水利固定资产投资 …………………………………………… 1
2 重点水利建设 ………………………………………………… 4
3 主要水利工程设施 …………………………………………… 8
4 水资源节约利用与保护 ……………………………………… 13
5 防洪抗旱 ……………………………………………………… 15
6 水利改革与管理 ……………………………………………… 17
7 水利行业状况 ………………………………………………… 24

Contents

I. Investment in Fixed Assets　28

II. Key Water Projects Construction　32

III. Key Water Facilities　35

IV. Water Resources Conservation, Utilization and Protection　40

V. Flood Control and Drought Relief　42

VI. Water Management and Reform　44

VII. Current Status of the Water Sector　52

2020年是全面建成小康社会目标实现之年，是全面打赢脱贫攻坚战和"十三五"规划收官之年，也是"两个一百年"奋斗目标的历史交汇之年。一年来，在习近平新时代中国特色社会主义思想指引下，各级水利部门认真贯彻落实党中央、国务院决策部署，落实"节水优先、空间均衡、系统治理、两手发力"的治水思路，积极践行"忠诚、干净、担当，科学、求实、创新"的新时代水利精神，面对新冠疫情、洪涝灾害的冲击，科学决策、迎难而上，真抓实干、砥砺奋进，各项水利工作取得明显成效。

1 水利固定资产投资

2020年,水利建设完成投资8181.7亿元,较上年增加1469.9亿元,增加21.9%。其中:建筑工程完成投资6014.9亿元,较上年增加20.6%;安装工程完成投资319.7亿元,较上年增加31.5%;机电设备及工器具购置完成投资250.0亿元,较上年增加13.1%;其他完成投资(包括移民征地补偿等)1597.1亿元,较上年增加26.8%。

	2013年/亿元	2014年/亿元	2015年/亿元	2016年/亿元	2017年/亿元	2018年/亿元	2019年/亿元	2020年/亿元	2020年比上年增加比例/%
全年完成	3757.6	4083.1	5452.2	6099.6	7132.4	6602.6	6711.7	8181.7	21.9
建筑工程	2782.8	3086.4	4150.8	4422.0	5069.7	4877.2	4987.9	6014.9	20.6
安装工程	173.6	185.0	228.8	254.5	265.8	280.9	243.1	319.7	31.5
设备及各类工器具购置	161.1	206.1	198.7	172.8	211.7	214.4	221.1	250.0	13.1
其他(包括移民征地补偿等)	640.2	605.6	873.9	1250.3	1585.2	1230.1	1259.7	1597.1	26.8

在全年完成投资中,防洪工程建设完成投资2801.8亿元,较上年增加22.4%;水资源工程建设完成投资3076.7亿元,较上年增加25.7%;水土保持及生态工程完成投资1220.9亿元,较上年增加

33.7%；水电、机构能力建设等专项工程完成投资1082.2亿元，较上年增加2.1%。

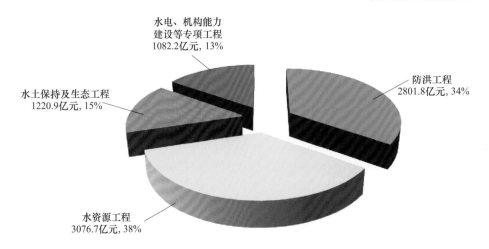

2020年分用途完成投资

七大江河流域完成投资6721.8亿元，东南诸河、西北诸河以及西南诸河等其他流域完成投资1459.9亿元；东部、中部、西部、东北地区完成投资分别为3511.2亿元、1955.6亿元、2446.5亿元和268.3亿元。

在全年完成投资中，中央项目完成投资48.5亿元，地方项目完成投资8133.1亿元。大中型项目完成投资1635.4亿元，小型及其他项目完成投资6546.2亿元。各类新建工程完成投资6009.1亿元，扩建、改建等项目完成投资2172.6亿元。

全年水利建设新增固定资产 4401.8 亿元。截至 2020 年年底,在建项目累计完成投资 19324.6 亿元,投资完成率为 60.9%;累计新增固定资产 9538.4 亿元,固定资产形成率为 49.4%,较上年减少 7.2 个百分点。

当年在建的水利建设项目 30005 个,在建项目投资总规模 31724.4 亿元,较上年增加 12.6%。其中:有中央投资的水利建设项目 16314 个,较上年减少 5.2%;在建投资规模 13308.9 亿元,较上年减少 5.3%。新开工项目 22532 个,较上年增加 8.4%,新增投资规模 7932.2 亿元,比上年增加 15.5%。全年水利建设完成土方、石方和混凝土方分别为 33.7 亿立方米、3.1 亿立方米、1.4 亿立方米。截至 2020 年年底,在建项目计划实物工程量完成率分别为:土方 96.3%、石方 75.2%、混凝土方 72.6%。

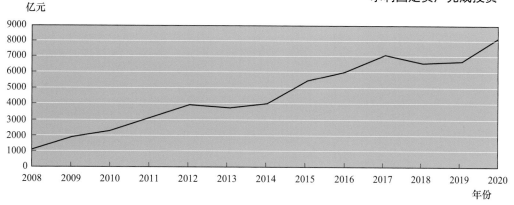

水利固定资产完成投资

2 重点水利建设

江河湖泊治理。2020年，在建江河治理工程5213处，其中：堤防建设585处，大江大河及重要支流治理860处，中小河流治理3224处，行蓄洪区安全建设及其他项目544处。截至2020年年底，在建项目累计完成投资4150.9亿元，项目投资完成率64.5%。长江中下游河势控制和河道整治工程有序实施；黄河下游防洪工程基本完工；进一步治淮38项工程已开工32项，其中13项建成并发挥效益；洞庭湖、鄱阳湖治理深入推进；太湖流域水环境综合治理21项工程已开工19项，其中17项已建成并发挥效益。

水库及枢纽工程建设。 2020年，在建水库及枢纽工程1315座。截至2020年年底，在建项目累计完成投资3648.7亿元，项目投资完成率为63.0%。青龙水库、溪头水库、板岭水库等23座中型水库开工。西藏湘河水利枢纽、青海那棱格勒河水利枢纽顺利截流；广西落久水利枢纽、河南前坪水库、重庆金佛山水库、安徽月潭水库等工程下闸蓄水；西江大藤峡水利枢纽实现蓄水、通航、发电的节点目标，左岸一期工程全面发挥综合效益。

三峡后续工作规划实施管理。 2020年，中央下达三峡后续工作计划投资83.65亿元，实际完成投资64.58亿元（不含自然资源部门项目，下同）；共安排各类项目1523个；项目开工率100%、完工率42.76%。全年共扶持特色产业发展项目305个，直接受益和就业移民群众4.1万人；实施教育培训与就业扶持项目353个，补助就读高职的三峡库区移民子女3465人，劳动力技能培训和就业扶持15354人；实施城镇移民小区帮扶项目156个、受益移民人数37.75万人，农村移民安置区帮扶项目610个、受益移民人数25.74万人；实施库岸环境综合整治长度45.17公里，系统治理支流11条；避险搬迁人数1877人；实施长江中下游影响区护岸长度58.97公里。

水资源配置工程建设。 2020年，水资源配置工程在建投资规模7365.2亿元，累计完成投资4188.7亿元，项目投资完成率为56.9%。重庆渝西水资源配置工程、新疆兵团奎屯河饮水工程等开工建设；引江济淮、滇中引水、引汉济渭、珠江三角洲水资源配置工程等加快实施。

农村水利建设。2020年，农村饮水安全巩固提升工程完成投资602.8亿元，其中中央补助资金15.0亿元。当年安排中央投资用于大中型灌区续建配套与节水改造70.8亿元，新建大型灌区工程42.1亿元，中型灌区节水改造60.9亿元。全年新增耕地灌溉面积870.4千公顷，新增节水灌溉面积1059.0千公顷，新增高效节水灌溉面积717.2千公顷。截至2020年年底，农村自来水普及率达到83%，农村集中供水率达到88%。

农村水电建设。2020年，全国农村水电建设完成投资58.6亿元，新增水电站69座，装机容量80.5万千瓦，其中：新投产装机57.4万千瓦，技改净增发电设备容量23.1万千瓦。

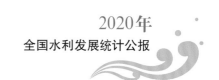

水土流失治理。2020年，水土保持及生态工程在建投资规模为3144.8亿元，累计完成投资1510.1亿元。全国新增水土流失综合治理面积6.4万平方公里，其中国家水土保持重点工程新增水土流失治理面积1.34万平方公里。对546座黄土高原淤地坝进行了除险加固。

行业能力建设。2020年，水利行业能力建设完成投资81.8亿元，其中：防汛通信设施投资9.5亿元，水文建设投资27.3亿元，科研教育设施投资1.7亿元，其他投资43.3亿元。

3 主要水利工程设施

堤防和水闸。截至 2020 年年底，全国已建成 5 级及以上江河堤防 32.8 万公里❶，累计达标堤防 24.0 万公里，堤防达标率为 73.0%，其中 1 级、2 级达标堤防长度为 3.7 万公里，达标率为 83.1%。全国已建成江河堤防保护人口 6.5 亿人，保护耕地 4.2 万千公顷。全国已建成流量为 5 立方米每秒及以上的水闸 103474 座，其中大型水闸 914 座。按水闸类型分，分洪闸 8249 座，排（退）水闸 18345 座，挡潮闸 5109 座，引水闸 13829 座，节制闸 57942 座。

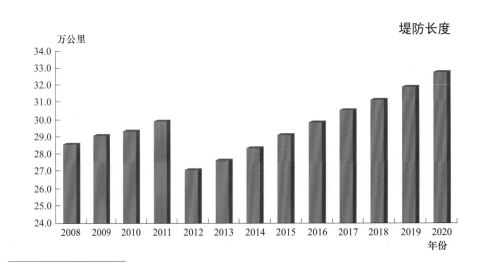

❶ 2011 年以前各年堤防长度含部分地区 5 级以下江河堤防长度。

水库和枢纽。全国已建成各类水库98566座，水库总库容9306亿立方米。其中：大型水库774座，总库容7410亿立方米；中型水库4098座，总库容1179亿立方米。

机电井和泵站。全国已累计建成日取水大于等于20立方米的供水机电井或内径大于等于200毫米的灌溉机电井共517.3万眼。全国已建成各类装机流量1立方米每秒或装机功率50千瓦以上的泵站95049处，其中：大型泵站420处，中型泵站4388处，小型泵站90241处。

灌区工程。全国设计灌溉面积2000亩及以上的灌区共22822处，耕地灌溉面积37940千公顷。其中：50万亩及以上灌区172处，耕地灌溉面积12344千公顷；30万~50万亩大型灌区282处，耕地灌溉面积5478千公顷。截至2020年年底，全国灌溉面积75687千公顷，耕地灌溉面积69161千公顷，占全国耕地面积的51.3%。全国节水灌溉工程面积37796千公顷，其中：喷灌、微灌面积11816千公顷，低压管灌面积11375千公顷。

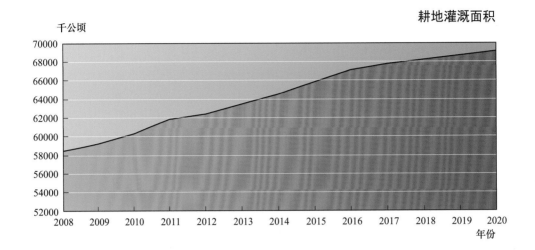

耕地灌溉面积

农村水电。截至 2020 年年底，全国已建成农村水电站 43957 座，装机容量 8133.8 万千瓦，占全国水电装机容量的 22.0%；全国农村水电年发电量 2423.7 亿千瓦时，占全国水电发电量的 17.9%。

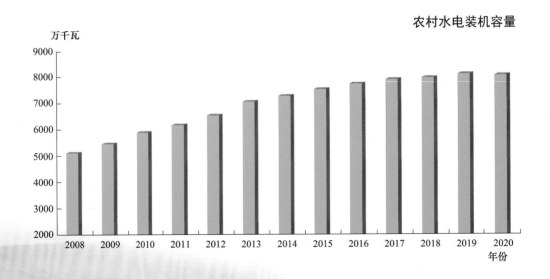

农村水电装机容量

水土保持工程。全国水土流失综合治理面积达 143.1 万平方公里❶，累计封禁治理保有面积达 21.4 万平方公里。2020 年持续开展全国全覆盖的水土流失动态监测工作，全面掌握县级以上行政区、重点区域、大江大河流域的水土流失动态变化。

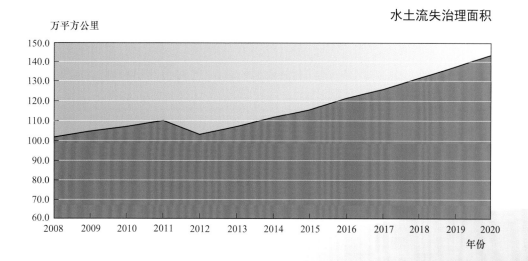

水文站网。全国已建成各类水文测站 119914 处，包括国家基本水文站 3265 处、专用水文站 4492 处、水位站 16068 处、雨量站 53392 处、蒸发站 8 处、地下水站 27448 处、水质站 10962 处、墒情站 4218 处、实验站 61 处。其中，向县级以上水行政主管部门报送水文信息的各类水文测站 71177 处，可发布预报站 2608 处，可发布预警站 2294 处；配备在线测流系统的水文测站 2066 处，配备视频监控系统的水文测站 4464 处。进一步完善了由中央、流域、省级和地市级共 336 个水质监测（分）中心和水质站（断面）组成的水质监测体系，监测范围覆盖

---

❶ 2012 年数据与第一次全国水利普查数据进行了衔接。

全国主要江河湖库和地下水、重要饮用水水源地、行政区界水域等。

水利网信。截至2020年年底，全国省级以上水利部门配置累计各类服务器9373台（套），形成存储能力47.0PB，存储各类信息资源总量达6.0PB，县级以上水利部门累计配置各类卫星设备3109台（套），利用北斗卫星短文传输报汛站达7297个，应急通信车65辆，集群通信终端3354个，宽、窄带单通信系统328套，无人机1338架。全国县级以上水利部门各类信息采集点达43.4万处，其中：水文、水资源、水土保持等各类采集点共约20.9万处，大中型水库安全监测采集点约22.5万处。

4 水资源节约利用与保护

水资源状况。2020年，全国水资源总量31605.2亿立方米，比多年平均值偏多14.0%；全国年平均降水量❶706.5毫米，比多年平均偏多10.0%，较上年增加8.5%。全国705座大型水库和3729座中型水库年末蓄水总量4358.7亿立方米，比年初减少237.5亿立方米。

水资源开发。2020年，新增规模以上水利工程❷供水能力104.8亿立方米。截至2020年年底，全国水利工程供水能力达8927.5亿立方米，其中：跨县级区域供水工程644.3亿立方米，水库工程2403.0亿立方米，河湖引水工程2138.1亿立方米，河湖泵站工程1815.1亿立方米，机电井工程1394.5亿立方米，塘坝窖池工程367.6亿立方米，非常规水资源利用工程164.9亿立方米。

---

❶ 2020年全国平均年降水量依据约18000个雨量站观测资料评价。
❷ 规模以上水利工程包括：总库容大于等于10万立方米的水库、装机流量大于等于1立方米每秒或装机容量大于等于50千瓦的河湖取水泵站、过闸流量大于等于1立方米每秒的河湖引水闸、井口井壁管内径大于等于200毫米的灌溉机电井和日供水量大于等于20立方米的机电井。

水资源利用。2020年,全年总供水量5812.9亿立方米,其中:地表水供水量4792.3亿立方米,地下水供水量892.5亿立方米,其他水源供水量128.1亿立方米。全国总用水量5812.9亿立方米,其中:生活用水863.1亿立方米,工业用水1030.4亿立方米,农业用水3612.4亿立方米,人工生态环境补水307.0亿立方米。与上年比较,用水量减少208.3亿立方米,其中:农业用水量减少69.9亿立方米,工业用水量减少187.2亿立方米,生活用水量减少8.6亿立方米,人工生态环境补水量增加57.4亿立方米。全国人均综合用水量为412立方米,农田灌溉水有效利用系数0.565,万元国内生产总值(当年价)用水量57.2立方米,万元工业增加值(当年价)用水量32.9立方米。按可比价计算,万元国内生产总值用水量和万元工业增加值用水量分别比2019年下降5.6%和17.4%。

5 防洪抗旱

2020年，全国洪涝灾害直接经济损失2669.8亿元（水利设施直接经济损失644.8亿元），占当年GDP的0.26%。全国农作物受灾面积7190千公顷，绝收面积1321.7千公顷，受灾7861.5万人次，因灾死亡失踪279人，倒塌房屋8.96万间[1]。安徽、四川、湖北、江西、重庆、湖南、甘肃等省（直辖市）受灾较重。

全国受旱地域分布较广，但造成的影响总体较轻，内蒙古、辽宁、浙江、福建、河南、广东、云南等省（自治区）旱灾比较严重。全国作物因旱受灾面积8352千公顷，成灾面积4081千公顷，直接经济总损失186亿元。全国因旱累计有669万城乡人口、449万头大牲畜发生临时性饮水困难。

---

[1] 2020年洪涝灾害直接经济损失、全国农作物受灾面积、绝收面积、受灾人口、因灾死亡和失踪人口、倒塌房屋数量等数据来源于应急管理部国家减灾中心。

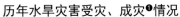

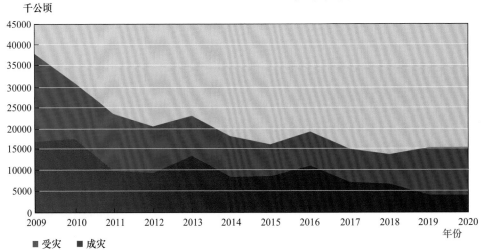

历年水旱灾害受灾、成灾[1]情况

全年中央下拨水利救灾资金28.5亿元，其中：防汛资金23.1亿元，抗旱资金5.4亿元。水利部组织指导各流域、各地科学调度水工程，其中调度大中型水库4042座次，共拦蓄洪水1780亿立方米；全国减淹城镇1334个次，减淹耕地约2277千公顷；避免转移人员约2213万人次。全年完成抗旱浇地面积15692千公顷，挽回粮食损失200亿公斤，解决了577万城乡居民和376万头大牲畜因旱临时饮水困难。

---

[1] 因机构改革后职能调整，水利部不再统计发布水灾成灾面积。图中成灾数据自2019年起不再包含水灾。

6 水利改革与管理

节约用水管理。2020年，国家节水行动重要节点目标指标全面完成，31个省（自治区、直辖市）全部出台省级实施方案。编制小麦等国家用水定额51项，印发34项，农业用水定额已覆盖88%粮食和85%油料作物播种面积，服务业、工业用水定额已分别覆盖行业用水总量90%和80%以上。建成298所节水型高校，推进43所高校实施合同节水，吸引社会资本超过1.6亿元，合同期内预计节水量5800万立方米。1790个市、县级水利机关及流域机构直属单位建成节水机关，平均节水率近30%。14493个工业、服务业、农业灌区用水单位纳入国家、省、市三级重点监控用水单位名录。2020年，全国非常规水源利用量达到128.1亿立方米，其中再生水利用量达到109.0亿立方米。

河（湖）长制。2020年，31个省（自治区、直辖市）党委和政府主要负责同志全部担任总河长，30多万名省、市、县、乡级河长湖长累计巡查河湖690万人次，协调解决河湖问题，督导落实"一河（湖）一策"方案。各地清理整治河湖"四乱"问题2.24万个，清除河道垃圾780万吨，拆除违法建筑810万平方米，清理非法占用河湖岸线

4400公里，拆除非法围堤830公里，查处非法采运砂船1900艘，河湖面貌明显改善。扎实推进长江干流岸线利用项目清理整治，共清理整治违法违规岸线利用项目2431个，腾退长江干流岸线158公里。强化监督检查，对31个省（自治区、直辖市）所有设区市的9164个河段（湖片）开展暗访督查；对长江、黄河、大运河等重点流域区域开展专项督查。

最严格水资源管理。水利部会同国家发展改革委等完成对31个省（自治区、直辖市）2020年度最严格水资源管理制度考核，浙江、江苏、山东、安徽等4个省考核等级为优秀。2020年，新批复了金沙江、沅江、西江等9条跨省江河水量分配，各省区批复109条跨地市江河水量分配。发布了两批共90条重点河湖、166个断面的生态流量目标。明确13个地市、62个县水资源超载，通过暂停其新增取水许可等政策措施加快推进超载综合治理。持续加强跨省江河流域水资源统一调度，2020年新启动实施了赤水河、牛栏江、北洛河等11条跨省江河水资源统一调度，黄河干流实现连续21年不断流，黑河流域下游东居延海连续16年不干涸。2020年，引黄入冀生态补水引黄河水14.53亿立方米，塔里木河流域向"四源一干"胡杨林区生态输水19.58亿立方米。中国水权交易所累计交易水量31.89亿立方米，交易金额21.13亿元，其中2020年全年共促成交易273宗，交易水量3亿立方米，交易金额3.64亿元。

运行管理。2020年，落实小型水库维修养护中央补助资金18亿元，带动地方财政投入13.3亿元。培训小型水库防汛"三个责任人"17万人。截至2020年年底，累计批准国家级水利风景区878个，其

中：水库型 373 个，自然河湖型 195 个，城市河湖型 195 个，湿地型 47 个，灌区型 31 个，水土保持型 37 个。

水价改革。配合国家发展改革委修订《水利工程供水价格管理办法》《水利工程供水定价成本监审办法》，积极推动水价改革。截至 2020 年年底，累计实施农业水价综合改革面积 2.9 亿亩，其中，2020 年新增农业水价综合改革面积 1.3 亿亩。

水利规划和前期工作。2020 年，中央层面审批（含印发审查意见）水利规划 20 项。全面启动"十四五"水安全保障规划编制工作，15 项专项规划（实施方案）中 3 项已印发、12 项已编制。组织开展国家水网工程规划纲要编制，全力抓好黄河流域生态保护和高质量发展、成渝地区双城经济圈建设、长江三角洲区域一体化发展、粤港澳大湾区建设等国家重大区域发展战略水利各项任务落实。加快推进重点流域和主要支流综合规划审批，批复沅江、雅砻江、赤水河、无定河、北洛河、郁江、诺敏河、绰尔河等流域综合规划。编制实施方案，谋划解决防汛中的薄弱环节，加快防洪工程建设。2020 年，国家发展改革委批复项目可行性研究报告 6 项，总投资 464.37 亿元；水利部批复初步设计 10 项，总投资 422.13 亿元。

水土保持管理。2020 年，水利部制定出台了生产建设项目水土保持承诺制管理、信用监管等 5 项制度。全国共审批生产建设项目水土保持方案 8.58 万个；完成水土保持设施自主验收的生产建设项目 2.22 万个。持续创新手段方式，首次实现生产建设项目水土保持遥感监管全国国土范围（除香港、澳门、台湾）全覆盖，通过卫星遥感解译和地方现场核查，共认定并查处"未批先建""未批先弃""超出防治责任范围"等违法违规项目 3.84 万个。

农村水电管理。截至 2020 年年底，23 个省份累计创建绿色小水电站 616 座。积极推进农村水电站安全生产标准化建设，全国累计建成安全生产标准化电站 2789 座，其中一级电站 82 座、二级电站 1168 座、三级电站 1539 座。全国共有 1283 条河流、1915 个生态改造项目、2071 个增效扩容项目完成改造，累计修复减脱水河段 3000 多公里。

水利移民。2020年，水利部实行移民搬迁的34项重大水利工程中，累计搬迁移民人口11.7万人，占搬迁移民总人口的59%；累计完成征地移民投资542亿元，占征地移民总投资的60%。国家核定当年新增大中型水库农村移民后期扶持人数14.6万人。

水利监督。2020年，组织开展了"2020年最严格水资源管理制度考核""水利行业监督检查"两大类36项工作内容，全年共派出检查组2851组次、13609人次，检查项目61327个，发现问题72919项；实施"约谈"及以上等级责任追究821家·次。水利行业共发生生产安全事故9起，死亡16人。开展水利行业安全生产专项整治三年行动，共辨识管控危险源近7.2万个，排查治理隐患5.4万余个。

依法行政。2020年，制定水利部规章2件。全国立案查处水事违法案件21031件，结案19935件，结案率94.8%；水利部共办结行政复议案件15件，另有国务院行政复议裁决案1件，办理行政应诉25件。

行政许可。2020年，水利部（包括部机关和各流域机构）共受理行政审批事项1387件，办结1327件。其中：水工程建设规划同意书审核41件，水利基建项目初步设计文件审批10件，取水许可260件，非防洪建设项目洪水影响评价报告审批31件，河道管理范围内建设项目工程建设方案审批386件，生产建设项目水土保持方案审批63件，国家基本水文测站设立和调整审批4件，专用水文测站审批1件，国家基本水文测站上下游建设影响水文监测工程审批 38 件，水利工程建设监理单位资质认定（新申请、增项、晋升）151 件，水利工程质量检测单位甲级资质认定（申报、延续）335 件。

水利科技。2020 年，围绕水利改革发展重点领域组织完成研究项目 21 个，启动实施流域水治理重大问题研究项目 9 项，编制"十四五"国家重点研发计划重大需求 16 项。联合国家自然科学基金委员会和中国长江三峡集团、国家电投集团设立长江水科学研究联合基金和黄河水科学研究联合基金，共落实研究经费 5 亿元，其中 2020 年已下达 7500 万元。发布《2020 年度成熟适用水利科技成果推广清单》，近百项成果得到广泛应用。组织实施水利技术示范项目 47 项，合计 3031.2 万元。完成水利科技成果登记 87 项，成果评价 43 项。改革重组 10 个部级重点实验室，1 家水利行业野外观测站列入国家野外站择优建设清单。

水利标准化。2020 年，发布水利技术标准 45 项，其中国家标准 3 项、行业标准 42 项，在编标准 143 项，废止标准 87 项。26 本小水电国际标准中文版获联合国工发组织与国际小水电联合会正式发布，《水库大坝安全评价导则》《农村饮水安全评价准则》分别获得"中国标准创新贡献奖"二等奖、三等奖，为时隔十余年再度获奖；《水工混凝土结构耐久性评定规范》《渡槽安全评价导则》分别获得"标准科技创新奖"一等奖、三等奖，为水利部首次获得。

国际合作。2020年，以部级领导互致信函、视频会议等方式组织召开多边、双边在线对外交流活动54场，参加外方主办在线活动112次。获得2024年第28届国际大坝会议主办权。拓展世界银行、亚洲开发银行渠道，立项实施项目3项，落实经费118万美元。稳步实施"一带一路"建设水利合作专项规划重点任务，启动2个援外项目，获批3个援外培训班，建立基于互联网+"水利工程医院"国际平台。帮助泰国应对湄公河旱情，将澜沧江水文信息共享范围由汛期扩大至全年，开通澜湄水资源合作信息共享平台网站，圆满完成对周边国家72个水文站国际报汛工作，跨界河流洪旱灾害防御、信息共享、联合研究、能力建设等领域互利共赢合作水平进一步提升。

7 水利行业状况

水利单位。截至2020年年底，水利系统内外各类县级及以上独立核算的法人单位22050个，从业人员89.9万人。其中：机关单位2654个，从业人员12.4万人，比上年减少0.1%；事业单位15205个，从业人员48.3万人，比上年减少3.5%；企业3530个，从业人员28.8万人，比上年减少6.1%；社团及其他组织661个，从业人员0.4万人，比上年减少35.3%。

职工与工资。全国水利系统从业人员 80.9 万人，比上年减少 5.27%。其中，全国水利系统在岗职工 77.8 万人，比上年减少 5.93%。在岗职工中，部直属单位在岗职工 6.7 万人，比上年增加 0.82%，地方水利系统在岗职工 71.1 万人，比上年减少 6.58%。全国水利系统在岗职工工资总额为 790.9 亿元，年平均工资 10.2 万元。

职工与工资情况

	2010年	2011年	2012年	2013年	2014年	2015年	2016年	2017年	2018年	2019年	2020年
在岗职工人数/万人	106.6	102.5	103.4	100.5	97.1	94.7	92.5	90.4	87.9	82.7	77.8
其中：部直属单位/万人	7.4	7.5	7.4	7.0	6.7	6.6	6.4	6.4	6.6	6.6	6.7
地方水利系统/万人	96.3	95.0	96.0	93.5	90.4	88.1	86.1	84.0	81.3	76.0	71.1
在岗职工工资/亿元	297.9	351.4	389.1	415.3	451.4	529.4	640.5	739.1	802.7	787.6	790.9
年平均工资/(元/人)	28816	34283	37692	41453	46569	55870	69377	83534	91307	95385	102000

全国水利发展主要指标（2015—2020年）

指标名称	单位	2015年	2016年	2017年	2018年	2019年	2020年
1. 灌溉面积	千公顷	72061	73177	73946	74542	75034	75687
2. 耕地灌溉面积	千公顷	65873	67141	67816	68272	68679	69161
其中：本年新增	千公顷	1798	1561	1070	828	780	870
3. 节水灌溉面积	千公顷	31060	32847	34319	36135	37059	37796
其中：高效节水灌溉面积	千公顷	17923	19405	20551	21903	22640	23190
4. 万亩以上灌区	处	7773	7806	7839	7881	7884	7713
其中：30万亩以上	处	456	458	458	461	460	454
万亩以上灌区耕地灌溉面积	千公顷	32302	33045	33262	33324	33501	33638
其中：30万亩以上	千公顷	17686	17765	17840	17799	17994	17822
5. 自来水普及率	%	76	79	80	81	82	83
农村集中供水率	%	82	84	85	86	87	88
6. 除涝面积	千公顷	22713	23067	23824	24262	24530	24586
7. 水土流失治理面积	万平方公里	115.5	120.4	125.8	131.5	137.3	143.1
其中：新增	万平方公里	5.4	5.6	5.9	6.4	6.7	6.4
8. 水库	座	97988	98460	98795	98822	98112	98566
其中：大型水库	座	707	720	732	736	744	774
中型水库	座	3844	3890	3934	3954	3978	4098
水库总库容	亿立方米	8581	8967	9035	8953	8983	9306
其中：大型水库	亿立方米	6812	7166	7210	7117	7150	7410
中型水库	亿立方米	1068	1096	1117	1126	1127	1179
9. 全年水利工程总供水量	亿立方米	6103	6040	6043	6016	6021	5813
10. 堤防长度	万公里	29.1	29.9	30.6	31.2	32.0	32.8
保护耕地	千公顷	40844	41087	40946	41409	41903	42168
堤防保护人口	万人	58608	59468	60557	62837	67204	64591
11. 水闸总计	座	103964	105283	103878	104403	103575	103474
其中：大型水闸	座	888	892	892	897	892	914

续表

指标名称	单位	2015年	2016年	2017年	2018年	2019年	2020年
12. 年末全国水电装机容量	万千瓦	31937	33153	34168	35226	35564	36972
全年发电量	亿千瓦时	11143	11815	11967	12329	12991	13540
13. 农村水电装机容量	万千瓦	7583	7791	7927	8044	8144	8134
全年发电量	亿千瓦时	2351	2682	2477	2346	2533	2424
14. 当年完成水利建设投资	亿元	5452.2	6099.6	7132.4	6602.6	6711.7	8181.7
按投资来源分：							
（1）中央政府投资	亿元	2231.2	1679.2	1757.1	1752.7	1751.1	1786.9
（2）地方政府投资	亿元	2554.6	2898.2	3578.2	3259.6	3487.9	4847.8
（3）国内贷款	亿元	338.6	879.6	925.8	752.5	636.3	614.0
（4）利用外资	亿元	7.6	7.0	8.0	4.9	5.7	10.7
（5）企业和私人投资	亿元	187.9	424.7	600.8	565.1	588.0	690.4
（6）债券	亿元	0.4	3.8	26.5	41.6	10.0	87.2
（7）其他投资	亿元	131.7	207.1	235.9	226.3	232.8	144.9
按投资用途分：							
（1）防洪工程	亿元	1930.3	2077.0	2438.8	2175.4	2289.8	2801.8
（2）水资源工程	亿元	2708.3	2585.2	2704.9	2550.0	2448.3	3076.7
（3）水土保持及生态建设	亿元	192.9	403.7	682.6	741.4	913.4	1220.9
（4）水电工程	亿元	152.1	166.6	145.8	121.0	106.7	92.4
（5）行业能力建设	亿元	29.2	56.9	31.5	47.0	63.4	85.2
（6）前期工作	亿元	101.9	174.0	181.2	132.0	132.7	157.3
（7）其他	亿元	337.5	636.2	947.5	835.8	757.4	747.3

说明：1. 本公报不包括香港特别行政区、澳门特别行政区及台湾省的数据。
2. 水利发展主要指标分别于2012年、2013年与第一次全国水利普查数据进行了衔接。
3. 农村水电的统计口径为单站装机容量5万千瓦及以下的水电站。

2020 STATISTIC BULLETIN ON CHINA WATER ACTIVITIES

Ministry of Water Resources, P. R. China

The year of 2020 witnesses the completion of building a moderately prosperous society in all respects. It is the closing year for winning the battle against poverty in an all-round way and the 13th Five-Year Plan, as well as the year when two centenary goals meet historically. Over the past year, under the guidance of Xi Jinping's thought of socialism with Chinese characteristics for a new era, water departments and authorities at all levels had made great efforts to implement all decisions and deployments of the CPC Central Committee and the State Council, followed the guidelines of "prioritize water saving, accomplish spatial equilibrium, implement systematic governance and achieve government-market synergy" and actively practiced the spirit of "loyalty, cleanliness, responsibility, respecting science, truth-seeking and innovation" for water governance in the new era, made scientific decisions to overcome difficulties brought by COVID-19 pandemic and flood disasters, worked hard and forged ahead, and made remarkable achievements in every respects of water management.

I. Investment in Fixed Assets

In 2020, the total investment in water projects amounted to 818.17 billion Yuan, with an increase of 146.99 billion or 21.9% compared with that in 2019, among which, 601.49 billion Yuan was being allocated for construction projects with an increase of 20.6%, 31.97 billion Yuan for installation with an increase of 31.5%, 25 billion Yuan for expenditure on purchases of machinery, electric equipment and instruments with an increase of 13.1%, and 159.71 billion Yuan for other purposes, including compensation for resettlement and land acquisition, with an increase of 26.8%.

	2013 /billion Yuan	2014 /billion Yuan	2015 /billion Yuan	2016 /billion Yuan	2017 /billion Yuan	2018 /billion Yuan	2019 /billion Yuan	2020 /billion Yuan	Increase /%
Total completed investment	375.76	408.31	545.22	609.96	713.24	660.26	671.17	818.17	21.9
Construction project	278.28	308.64	415.08	442.20	506.97	487.72	498.79	601.49	20.6
Installation project	17.36	18.50	22.88	25.45	26.58	28.09	24.31	31.97	31.5
Purchase of machinery, equipment and instruments	16.11	20.61	19.87	17.28	21.17	21.44	22.11	25.00	13.1
Others (including compensation of resettlement and land acquisition)	64.02	60.56	87.39	125.03	158.52	123.01	125.97	159.71	26.8

In the total completed investment, 280.18 billion Yuan was allocated to the construction of flood control projects, up 22.4% from the previous year; 307.67 billion Yuan was for the construction of water resources projects, up 25.7%; 122.09 billion Yuan was for soil and water conservation and ecological restoration, up 33.7%; and 108.22 billion Yuan for specific projects of hydropower development and capacity building, up 2.1% over the previous year.

The competed investment for seven major river basins reached 672.18 billion Yuan, of which 145.99 billion Yuan was invested in river basins in the southeast, southwest and northwest of China, while investments in river basins in east, central, west and northeast China were 351.12 billion Yuan, 195.56 billion Yuan, 244.65 billion Yuan and 26.83 billion Yuan, respectively.

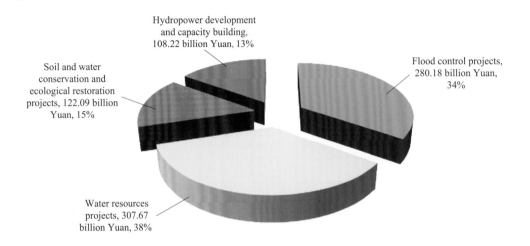

Completed investment of projects in 2020

Of the total competed investment, the Central Government contributed 4.85 billion Yuan, and local governments contributed 813.31 billion Yuan. Investments for large and medium-sized projects were 163.54 billion Yuan; and for small and other projects were 654.62 billion Yuan. Investments for new projects and rehabilitation and expansion projects were 600.91 billion Yuan and 217.26 billion Yuan.

The newly-added fixed asset of the year in water project construction totaled 440.18 billion Yuan. By the end of 2020, the accumulated investment in projects under construction was 1,932.46 billion Yuan, with the completion rate reaching 60.9%. New fixed assets totaled 953.84 billion Yuan and the rate of investment transferred into fixed assets was 49.4%, a decrease of 7.2 percentage point over the previous year.

A total of 30,005 water projects were under construction in 2020, with a total investment of 3,172.44 billion Yuan, an increase of 12.6% over the previous year, among which, 16,314 projects were funded by the Central Government, a decrease of 5.2% over the previous year, occupying 1,330.89 billion Yuan, down 5.3% over the previous year. There were 22,532 projects launched in 2020, an increase of 8.4%, with an increase of investment totaled 793.22 billion Yuan, an increase of 15.5% over the previous year. The completed civil works of earth, stone and concrete structures were 3.37 billion m^3, 310 million m^3, and 140 million m^3, respectively. By the end of 2020, the rate of completed quantity of earthwork, stonework, and concrete of under-construction projects were 96.3%, 75.2% and 72.6%, respectively.

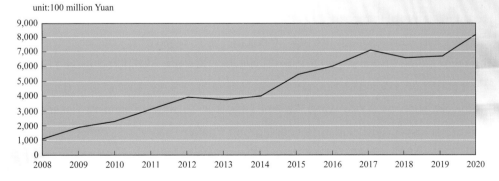

Completed investment for fixed water assats

II. Key Water Projects Construction

Harness of large rivers and lakes. In 2020, there were 5,213 river harness projects under construction, including 585 flood control dyke and embankment construction projects, 860 projects for large river and main tributary control, and 3,224 medium-sized and small river control works, and 544 flood diversion and storage area construction or other projects. By the end of 2020, the accumulated investment in projects under construction was 415.09 billion Yuan, with a completion rate of 64.5%. The projects for river regime control and river course training and restoration in the middle and lower reaches of the Yangtze River had been implemented in an appropriate manner. Flood control works in the lower reaches of the Yellow River were mostly completed. Out of the 38 Huaihe River improvement projects, 32 started construction, among which 13 were put into operation for benefit generation. Improvement work for Dongting and Poyang lakes also made significant progress. Up to 19 out of the 21 projects for the Comprehensive Improvement of Water Environment of Taihu Lake began construction, among which 17 projects completed construction with benefit generation.

Reservoir and water control projects. In 2020, there were 1,315 reservoir projects under construction. By the end of the year, completed investment of under-construction projects reached 364.87 billion Yuan, with a completion rate of 63.0%. Construction commenced for 23 medium-sized reservoirs, namely Qinglong Reservoir, Xitou Reservoir and Banling Reservoir. Xianghe Multipurpose Dam Project in Tibet and Nalinggele Multipurpose Dam Project in Qinghai realized river damming. Reservoirs such as Luojiu Water Control Project in Guangxi, Qianping Reservoir in Henan, Jinfoshan Reservoir in Chongqing and Yuetan Reservoir in Anhui completed impoundment. Datengxia Multipurpose Dam Project in Xijiang River Basin has achieved the milestones of water storage, navigation and power generation, and phase-I of the Left Bank Project has started to generate comprehensive benefits.

Implementation and management of the follow-up working planning of the Three Gorges Project. In 2020, the Central Government made an investment plan of 8.365 billion Yuan for the follow-up work of the Three Gorges Project, which has completed 6.458 billion Yuan (excluding projects with investments from the Ministry of Natural Resources, the same below). A total of 1,523 projects of various types were planned, with an operating rate of 100% and a completion rate of 42.76%. The funds had assisted 305 special-featured industry development projects, with 41,000 resettled persons benefited and employed. The funds assisted 353 education, training and employment support projects, with subsidence to 3,465 children of resettled families in the Three Gorges Reservoir Area, to attend higher vocational colleges. There were 15,354 persons attend labor skills and employment training. There were 156 assistance projects of urban immigrant community implemented for benefiting 377,500 immigrants. There were 610 assistance projects of rural immigrant resettlement area for benefiting 257,400 immigrants. There were 45.17 kilometers of reservoir banks were restored for environment improvement, and water systems 11 tributaries were improved. There were 1,877 people relocated for risk mitigation. The length of revetment for the affected area in the middle and lower reaches of the Yangtze River was 58.97 kilometers.

Water allocation projects. In 2020, investment for water allocation projects under construction reached 736.52 billion Yuan and completed investment accumulated to 418.87 billion Yuan, accounting for 56.9% of the total. The projects of Yuxi Water Allocation in Chongqing and Water diversion from Kuitun River of the Xinjiang Production and Construction Corps started construction. Progressed sped up for the projects of water diversion from the Yangtze River to the Huaihe River, water diversion in central Yunnan, water diversion from the Hanshui River to the Weihe River in Shaanxi, and water allocation in the Pearl River Delta.

Rural water conservancy construction. In 2020, completed investment for strengthening and improving safe drinking water supply in rural areas reached 60.28 billion Yuan, including 1.5 billion Yuan of central government subsidies. The Central Government allocated 7.08 billion Yuan for the construction of large and medium irrigation and drainage systems and rehabilitation of irrigation districts for water saving purpose, 4.21 billion Yuan for new large-scale irrigation district projects and 6.09 billion Yuan for water-saving renovation of medium-sized irrigation districts. The newly-added irrigated area was 870,400 ha. Water-saving irrigated area increased by 1,059,000 ha, and area covered by highly-efficient water-saving irrigation grew by 717,200 ha. By the end of 2020, the rural population access to tap water made up a percentage of 83% and the coverage of centralized water supply raised to 88%.

Rural hydropower and electrification. In 2020, completed investment of rural hydropower station construction nationwide amounted to 5.86 billion Yuan, adding 69 new hydropower stations, with a total installed capacity of 0.805 million kW. Of these, newly installed capacity amounted to 0.574 million kW, and capacity increase resulted from technical rehabilitation was 0.231 million kW.

Soil and water conservation. In 2020, a total of 314.48 billion Yuan was allocated to the under-constructed projects for soil and water conservation and ecological restoration, with an accumulated investment of 151.01 billion Yuan. The newly-added areas for comprehensive control of soil erosion reached 64,000 km^2, of which the areas under the National Major Project for Soil Conservation was 13,400 km^2. Up to 546 silt-retention dams on Loess Plateau at high risk were strengthened and rehabilitated.

Capacity building. The completed investment for capacity building in 2020 was 8.18 billion Yuan, of which 950 million Yuan was spent on communication equipment for flood control, 2.73 billion Yuan on hydrological facilities, 170 million Yuan for scientific research and education facilities and 4.33 billion Yuan for others.

III. Key Water Facilities

Embankments and water gates. By the end of 2020, the completed river dykes and embankments at Grade-V or above had a total length of 328,000 km❶. The accumulated length of dykes and embankments that met the standard reached 240,000 km, accounting for 73.0% of the total. Among which, Grade-I and Grade-II dykes and embankments were up to the standard reached 37,000 km, or 83.1% of the total. All completed dykes and embankments nationwide can protect 650 million people and 42,000 ha of cultivated land. The number of completed water gates with a flow of 5 m³/s increased to 103,474, of which 914 were large water gates. By type, there were 8,249 flood diversion sluices, 18,345 drainage/return water sluices, 5,109 tidal barrages, 13,829 water diversion intakes and 57,942 controlling gates.

---

❶ The length of river dykes in each year before 2011 included the length of river dykes below Grade-V in some areas.

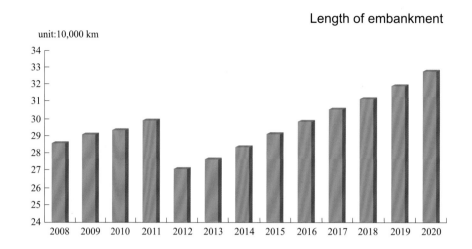

Reservoirs and water complexes. The number of completed reservoirs in China reached 98,566, with a total storage capacity of 930.6 billion m^3. Of which 774 reservoirs are large reservoirs, with a total capacity of 741.0 billion m^3 and 4,098 reservoirs are medium-sized with a total capacity of 117.9 billion m^3.

Tube wells and pumping stations. Accumulatively, a total of 5.173 million tube wells, with a daily water abstraction capacity equal to or larger than 20 m^3 or an inner diameter equal to or larger than 200 mm, had been completed for water supply in the whole country. A total of 95,049 pumping stations with a flow of 1 m^3/s or an installed voltage above 50 kW were put into operation, including 420 large, 4,388 medium and 90,241 small pumping stations.

Irrigation systems. The irrigation districts with a designed area of 2,000 mu or above were 22,822 in total, covering 37.940 million ha of irrigated farmland. Of which, 172 irrigation districts had an irrigated area of 500,000 mu or above, and their total irrigated area reached 12.344 million ha. The irrigation districts with an area from 300,000 to 500,000 mu were 282, covering 5.478 million ha of irrigated farmland. By the end of 2020, the total irrigated area amounted to 75.687 million ha. The irrigated area of cultivated land reached 69.161 million ha that accounted for 51.3% of the total in China. The areas with water-saving irrigation facilities totaled 37.796 million ha, among which 11.816 million ha were equipped with sprinklers or micro-irrigation systems and 11.375 million ha with low-pressure pipes.

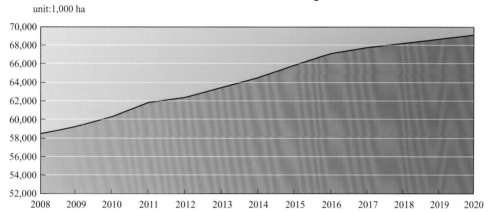

Rural hydropower and electrification. By the end of 2020, hydropower stations built in rural areas totaled 43,957, with an installed capacity of 81.338 million kW, accounting for 22.0% of the national total. The annual power generation by these hydropower stations reached 242.37 billion kW·h, accounting for 17.9% of the national total.

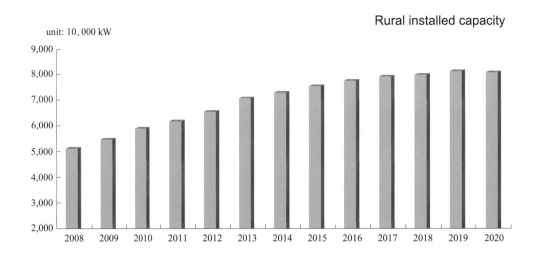

Soil and water conservation. By the end of 2020, the restored eroded areas reached 1.431 million km² ❶; and the forbidden area for ecological restoration accumulated to 214,000 km². Dynamic monitoring for soil and water loss had been continued in 2020 in all administrative areas above county level, key areas, and major river basins in the country, to gain a comprehensive understanding of dynamic changes.

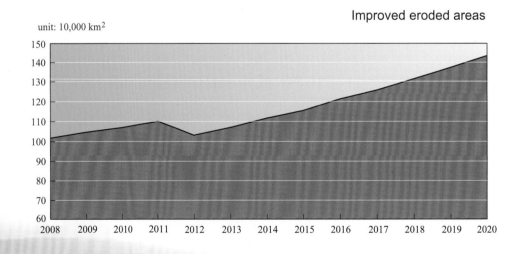

---

❶ The data in 2012 were linked to the data of the First National Water Conservancy Survey.

Hydrological station networks. In 2020, the number of hydrological stations of all kinds totaled 119,914 in the whole country, including 3,265 national basic hydrologic stations, 4,492 special hydrologic stations, 16,068 gauging stations, 53,392 precipitation stations, 8 evaporation stations, 27,448 groundwater monitoring stations, 10,962 water quality stations, 4,218 soil moisture monitoring stations and 61 experimental stations. Among them, 71,177 stations of various kinds can provide hydrological information to water administration authorities at and above county level; 2,608 stations can provide forecasting and 2,294 may issue early warnings; 2,066 were equipped with online flow measurement and 4,464 were equipped with video monitors. A water quality monitoring system, including 336 monitoring centers and sub-centers as well as water quality stations (sections) at central, basin, provincial and local levels, had been formed, and its scope covers major rivers, lakes and reservoirs, groundwater, important drinking water sources, and water bodies in administrative boundaries.

Water networks and information systems. By the end of 2020, the water resources departments and authorities at and above provincial level were equipped with 9,373 servers of varied kinds, forming a total storage capacity of 47.0 PB, and keeping 6.0 PB of data and information. The water resources departments and authorities at and above county level had equipped with 3,109 sets of various kinds of satellite equipment, 7,297 flood forecasting stations for short message transmission from the Beidou Satellites, 65 vehicles for emergency communication, 3,354 cluster communication terminals, 328 narrowband and broadband communication systems, and 1,338 unmanned aerial vehicles (UAV). A total of 434,000 information gathering points were available for water resources departments and authorities at and above county level, including 209,000 points for collecting data of hydrology, water resources and soil and water conservation and 225,000 points for safety monitoring at large and medium-sized reservoirs.

IV. Water Resources Conservation, Utilization and Protection

Water resources conditions. The total national water resources in 2020 was 3,160.52 billion m^3, approximately 14.0% more than the normal years. The mean annual precipitation❶ was 706.5 mm, 10.0% more than normal years and 8.5% more than the previous year. By the end of 2020, the total storage of 705 large and 3,729 medium-sized reservoirs were 435.87 billion m^3, 23.75 billion m^3 less than that at the beginning of the year.

Water resources development. In 2020, the newly-increased water supply capacity by water facilities above designated size❷ was 10.48 billion m^3. By the end of 2020, the total water supply capacity of China reached 892.75 billion m^3, including 64.43 billion m^3 of water diverted from cross-county facilities, 240.30 billion m^3 from reservoirs, 213.81 billion m^3 from river and lake diversion schemes, 181.51 billion m^3 from pumping stations along rivers and lakes, 139.45 billion m^3 from tube wells, 36.76 billion m^3 from ponds, weirs and cellars, and 16.49 billion m^3 from unconventional water sources.

Water resources utilization. In 2020, the total quantity of water supply amounted to 581.29 billion m^3, including 479.23 billion m^3 from surface water, 89.25 billion m^3 from groundwater and 12.81 billion m^3 from other sources. The total water consumption amounted to 581.29 billion m^3, among which domestic water use amounted to 86.31 billion m^3, industrial water use totaled 103.04 billion m^3,

---

❶ Average precipitation of 2020 was based on data from approximately 18,000 stations.

❷ Water projects above designated size include reservoirs with a total capacity of 100,000 m^3 or higher, pump stations with an installed flow at or above 1 m^3/s or an installed capacity at or above 50 kW, water diversion gates with a flow at or above 1 m^3/s, electric irrigation wells 200 mm or larger in inner diameter or with a water supply capacity at or above 20 m^3 per day.

agricultural water use was 361.24 billion m^3, artificial recharge for environmental and ecological use 30.70 billion m^3. Comparing to the previous year, the total water consumption decreased by 20.83 billion m^3. Agricultural water use decreased by 6.99 billion m^3, industrial water use decreased by 18.72 billion m^3, domestic water use decreased by 860 million m^3 while artificial recharge for environmental and ecological purposes increased by 5.74 billion m^3. Water consumption per capita in 2020 was 412 m^3 in average. The coefficient of effective irrigated water use was 0.565. Water use per 10,000 Yuan of GDP (at comparable price of the same year) was 57.2 m^3 and that per 10,000 Yuan of industrial value added (at comparable price of the same year) was 32.9 m^3. Based on estimation at comparable prices, water uses per 10,000 Yuan of GDP and per 10,000 Yuan of industrial value added decreased by 5.6% and 17.4% over the previous year respectively.

V. Flood Control and Drought Relief

In 2020, the direct economic loss of flood disasters was 266.98 billion Yuan (including 64.48 billion Yuan of direct losses as result of water facility damage), accounting for 0.26% of GDP in the same year. A total of 7,190,000 ha of cultivated land were affected by floods, 1,321,700 ha of farmland had no harvest, 78.615 million people were affected with 279 dead and missing. A total of 89,600 houses collapsed❶. Provinces suffered heavily from severe flooding included Anhui, Sichuan, Hubei, Jiangxi, Chongqing, Hunan, Gansu and other provinces (municipalities directly under the Central Government).

Drought was widespread but not severe on the whole. The seriously affected provinces and autonomous regions were Inner Mongolia, Liaoning, Zhejiang, Fujian, Henan, Guangdong, Yunnan and other provinces. The affected area of farmland was 8,352,000 ha and areas with no harvest❷ reached 4,081,000 ha, with a total of 18.6 billion Yuan of direct economic loss. A total of 6.69 million urban and rural residents and 4.49 million man-feed big animals and livestock suffered from temporary drinking water shortage.

---

❶ The data of direct economic losses caused by floods in 2020, the affected area of crops in China, the area of no harvest, the affected population, the number of dead and missing persons due to disasters, and the number of collapsed houses come from the National Disaster Reduction Center of China.

❷ Due to the adjustment of functions after the institutional reform, the Ministry of Water Resources no longer publishes statistics on flood-stricken areas. The disaster data in the figure no longer includes floods since 2019.

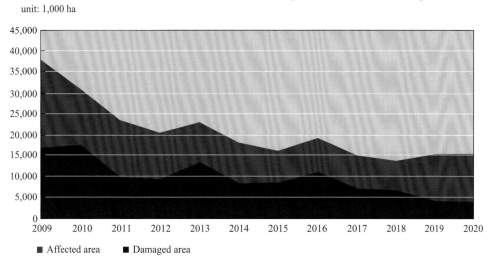

Flood or drought affected and damaged areas

In 2020, the Central Government allocated a total of 2.85 billion Yuan for water-related disaster mitigation, including 2.31 billion Yuan for flood defense and 540 million Yuan for drought relief. The Ministry of Water Resources (MWR) had provided guidance to relevant authorities in making scientific dispatching of water structures in various river basins and locations. There were 4,042 large and medium-sized reservoirs dispatched to store floodwater of 178 billion m^3, which prevented flooding of 1,334 times in cities and towns, protected 2,277 thousand ha of cultivated land, and about 22.13 million people being migrated. Up to 15,692,000 ha of drought-hit cultivated land were irrigated, reducing grain loss of 20 billion kg. Drinking water was provided to 5.77 million rural and urban population and 3.76 million big animals and livestock in order to alleviate temporary water shortage.

VI. Water Management and Reform

Water conservation management. In 2020, key target indicators for national water-saving actions were all completed. Detailed implementation plans were formulated by 31 provinces (autonomous regions and municipalities directly under the Central government). There were 51 national water quotas for wheat and others compiled and 34 issued. The quotas for agriculture water use have covered 88% of harvested area of grain and 85% of oil crops, and quotas for service sector and industrial water use have covered 90% and 80% of the total consumption respectively. There were 298 universities listed as water-saving type. Contracts were signed by 43 universities for water conservation that attracted more than 160 million Yuan of social capital, in order to save 58 million m^3 of water during the contract period. There were 1,790 municipal and county-level water departments or organizations directly under river basin authorities listed as water-saving type, who should meet the average water-saving rate of nearly 30%. There were 14,493 companies or units in industry, service sector and irrigation districts for agricultural production listed in the name-list of key monitored water users at national, provincial and municipal levels. In 2020, consumption of unconventional water sources in China reached 12.81 billion m^3, of which the consumption of recycled water reached 10.90 billion m^3.

River (lake) chief system. In 2020, all of the 31 provinces, autonomous regions and municipalities had CPC and government leaders serving as river (lake) chiefs. More than 300,000 river (lake) chiefs at province, city, county and town levels made a total of 6.90 million times of inspections for rivers and lakes and implementation of "one policy for one river/lake". A special action was initiated to address the issues such as misappropriation, illegal sand excavation, disposing of wastes and building structures without permission. A total of 22,400 illegal activities were reported and corrected, with 7.8 million tons of garbage cleaned up in river courses, 8.1 million km^2 of illegal structures dismantled, 4,400 km of banks freed

from illegal use, 830 km of enclosed embankments removed, and 1,900 illegal sand vessels banned, which resulted in great improvement in rivers and lakes. In addition, a clean-up action was conducted to shorelines of the Yangtze River, identifying and rectifying 2,431 suspected illegal structures and sties, and vacating 158 km of the shorelines of the main stream of the Yangtze River. Supervision and inspection were strengthened, as 9,164 river sections (lakes) in all cities with districts under their jurisdiction in 31 provinces, autonomous regions and municipalities were inspected without prior notice. Special investigations or inspections were made to oversee key river basins such as the Yangtze River, the Yellow River and the Grand Canal.

Most stringent water resources management. In 2020, MWR, in collaboration with the National Development and Reform Commission (NDRC) and other ministries, completed performance evaluation for the implementation of the most stringent water resources management system in 31 provinces, autonomous regions and municipalities in 2020, 4 provinces including Zhejiang, Jiangsu, Shanghai and Anhui were evaluated as excellent. Water allocation plans of 9 cross-provincial river basins, including Jinsha River, Yuanjiang River and Xijiang River, were approved. There were 109 water allocation plans for river basins across cities or counties approved by agencies at provincial or autonomous region level. Targets of ecological flow were defined for two batches of 90 key rivers and lakes and 166 sections. There were 13 cities or prefecture and 62 counties were identified as overuse of water, based on water availability, and asked to stop issue of new permits and take further control measures. Integrated water regulation and dispatching in inter-provincial basins was further strengthened, as 11 inter-provincial rivers, including Chishui River, Niulan River and Beiluo River, were integrated for unified regulated in 2020. The mainstream of the Yellow River had kept continuous flow for 21 consecutive years without dry up in the downstream. The East Dongjuyan Lake in the lower reaches of the Heihe River had been prevented from drying up for 16 consecutive years. In 2020, 1.453 billion m^3 of the Yellow River water was diverted into Hebei for ecological purpose, and 1.958 billion m^3 of environmental flow was released from Tarim River to the Populus euphratica forest areas in the basin. China Water Exchange has completed a total trading of 3.189 billion m^3 of water, with a value of 2.113 billion Yuan. In which, 273 times of entitlement trading was completed in 2020, with an amount of 300 million m^3 of water and a value of 364 million Yuan.

Operation and management. In 2020, the central government allocated 1.8 billion Yuan as subsidies for small reservoir repair and maintenance, and local governments allocated 1.33 billion Yuan as counterpart funding. The number of management staff trained for operation and maintenance of small reservoirs was up to 170,000. By the end of 2020, the number of approved national water scenic spots reached 878, including 373 reservoirs, 195 natural rivers and lakes, 195 lake or riverine cities, 47 wetlands, 31 irrigation districts and 37 soil conservation areas.

Water pricing reform. In collaboration with the NDRC, MWR revised *the Measures for the Administration of Water Supply Price for Water Resources Projects* and *the Measures for Supervision and Examination of Pricing and Cost of Water Supply for Water Resources Projects*, in order to promote water pricing reform. By the end of 2020, the area implemented reform of agricultural water pricing accumulated to 290 million mu, with an increase of 130 million mu in 2020.

Water resources planning and early-stage work. In 2020, there were 20 water resources plans approved by central government authorities (including printed and issued review comments). The drafting of Water Security Plan of 14th Five-Year Plan Period was fully launched. Out of 15 special plans/implementation plans, there were 3 planed made public and 12 completed drafting. The plan outline of national water network project was completed. The main tasks for implementing national strategy of regional development were fully committed, including ecological protection and high-quality development of the Yellow River Basin, Chengdu-Chongqing Economic Circle, regional integration development of the Yangtze River Delta, and development of Guangdong-Hong Kong- Macao Greater Bay Area. The process of examination and approval was accelerated for overall planning of key river basins and major tributaries, with approval of plans for Yuanjiang River, Yalong River, Chishui River, Wuding River, Beiluo River, Yujiang River, Nuomin River and Chuer River. Implementation plan was prepared especially for the weak links in flood control, so as to accelerate construction of flood control projects. In 2020, a total of 6 feasibility study reports were approved by NDRC, with a total investment of 46.437 billion Yuan. Up to 10 preliminary designs were approved by MWR, with a total investment of 42.213 billion Yuan.

Soil and water conservation. In 2020, MWR formulated and issued 5 systems, such as commitment and management system for soil and water conservation and credit supervision for production and construction projects. A total of 85,800 soil and water conservation plans of construction projects were approved. Water and soil conservation facilities of 22,200 construction projects completed self-check and acceptance. Innovative measures were continuously adopted with the full coverage of remote sensing supervision of soil and water conservation in production and construction projects (except Hong Kong, Macao and Taiwan) for the first time. According to on-site review based on interpretation of remote sensing satellite, 38,400 projects were investigated and punished because of construction without approval, damping wastes or failing to comply with the scope of responsibility for prevention and control.

Rural hydropower management. By the end of 2020, the title of green small hydropower station was awarded to 616 projects in 23 provinces. Standards for safe production and operation had been applied to hydropower stations in rural areas. The completed hydropower stations that complied with relevant standards accumulated to 2,789 in the whole country, including 82 level one, 1,168 level two and 1,539 level three hydropower stations. Rehabilitation was completed for 1,283 rivers, 1,915 ecological restoration projects and 2,071 expansion projects for efficiency increase. Over 3,000 km of river courses were recovered from low water level or drying up.

Resettlement of water projects. In 34 major water resources projects under the MWR, the resettlement was 117,000 in 2020, accounting for 59% of the total resettled population. Accumulated investment in land acquisition and resettlement is 54.2 billion Yuan, accounting for 60% of the total. According to the data approved by the central government, the relocated people in rural areas due to the construction of large and medium-sized reservoirs in 2020 were totaled 146,000,

who shall be given later support.

Safely Supervision. In 2020, there were 36 tasks of two categories were organized and implemented, including the Most Stringent Water Resources Management System in 2020 and supervision and inspection of the water sector. A total of 2,851 inspection teams with 13,609 person-times were dispatched for 61,327 projects, identifying 72,919 problems. Up to 821 "discipline conversations" or higher-level punishments were conducted. 9 production accidents happened with 16 people dead. The action of three-year inspections was initiated by MWR for safe production of the water sector, which identified 72,000 potential hazards with 54,000 properly handled.

Legislation and administrative law enforcement. In 2020, 2 regulations documents were formulated. In 2020, the number of water-related cases investigated totaled 21,031 and 19,935 cases or 94.8% were resolved. MWR handled and concluded 15 administrative reconsideration cases, 1 administrative reconsideration decision of the State Council and 25 administrative proceedings.

Administrative permits. In 2020, MWR handled 1,387 applications for water-related administrative approvals or permits with 1,327 completed, including 41 project plan approvals, 10 preliminary design reports of water construction projects, 260 water abstraction licenses, 31 evaluation reports of flood impact by non-flood control project, 386 plans of construction projects within the jurisdiction of river courses, 63 approvals of soil and water conservation plan of production and construction projects, 4 approvals for establishment and adjustment of national basic hydrological stations, 1 approval of special hydrological station, 38 approvals of hydrological monitoring projects for evaluating impact of construction at upper and lower of the national basic hydrological stations, 151 qualification approvals (including new application, adding of new items or promotion) for construction supervisors of water resources projects; and 335 Grade-A qualification identifications (including new application and extension) for quality supervisors of water-related projects.

Water science and technology. In 2020, there were 21 research projects organized and completed, which focused on key areas of water reform and development. 9 main subject studies on water governance of river basins were launched, and 16 key requirements in the plan for national key research and development during the 14th Five-Year Plan were proposed. Joint funds for scientific studies on the Yangtze River and the Yellow River, founded by MWR, National Natural Science Foundation of China, China Three Gorges Corporation and State Power Investment, raised 500 million Yuan, of which 75 million was disbursed in 2020. The Promotion List of Mature and Applicable Water Scientific and Technological Achievements in 2020 was released, with nearly 100 achievements widely applied and extended. There were 47 water technology demonstration projects implemented, with 30.312 million Yuan of funding. There were 87 water scientific and technological achievements registered and 43 of which evaluated. Restructuring of currently existed 10 ministerial-level key laboratories were completed. 1 field observation station of the water sector was listed in the national field stations for preferred construction.

Water-related Norms and Standards. In 2020, there were 45 water-related technical standards issued, including 3 national standards, 42 sectoral standards. The standards under permanent use were 143 and 87 abolished. The Chinese version of 26 international standards for small hydropower station was officially released by UNIDO and the International Network on Small Hydro Power. *The Guidelines on Dam Safety Evaluation* and *the Guidelines on Evaluation Criteria for Rural Drinking Water Safety* won the second and third prizes respectively of China

Standard Innovation and Contribution Prizes. In addition, *the Specifications for Evaluating Durability Evaluation of Hydraulic Concrete Structures* and *the Guidelines for Evaluation of Aqueduct Safety* won the first and third prizes respectively of the Awards for Standard of Science and Technology Innovation. As MWR won the China National Standards Innovation and Contribution Award again after more than ten years, and the Awards for Standard of Science and Technology Innovation for the first time, the competitiveness and influence of these technical standards had further strengthened.

International cooperation. In 2020, MWR organized 54 multi-bilateral online foreign exchange activities by means of exchange of letters among ministerial leaders and video conferences. There were 112 online activities sponsored by foreign parties were participated. MWR won the right to host the 28th International Conference on Dams in 2024. 3 projects were approved or implemented with 1.18 million US dollars of funding through the financial channel of the World Bank and the Asian Development Bank. Key tasks in the special plan of the Belt and Road initiative for water cooperation were implemented effectively, with two China's Foreign Aid projects initiated, three training courses for Foreign Aid projects approved. The international platform based on Internet + "water project hospital" was set up. China provided assistance to Thailand for drought relief by expanding the scope of Lancang River hydrological information sharing from flood season to the whole year. The website for platform of Lancang-Mekong Water Resources Cooperation Information Sharing was opened. Flood warning and reporting was conducted for 72 hydrological stations of the neighboring countries. Win-win cooperation for mutual benefits of the riparian countries in the transboundary river basins has been enhanced in the respects of flood and drought disaster prevention, information sharing, joint research and capacity building.

VII. Current Status of the Water Sector

Water-related institutions. By the end of 2020, there were 22,050 legal entities, with 899 thousands employees and separate accounts, engaged in water-related activities within the administrative jurisdiction at county level or above. Among them, the number of governmental organizations were 2,654 with 124 thousands employees, down by 0.1% over the previous year; public organizations were 15,205 with 483 thousands employees, down by 3.5%; enterprises were 3,530 with 288 thousands employees, down by 6.1%; societies and other institutions were 661 with 4 thousands employees, down by 35.3%.

Employees and salaries. Employees of the water sector totaled 809 thousands, down 5.27% from the previous year. Of which, in-service staff members amounted to 778 thousands, down 5.93%, including 67 thousands working in agencies directly under the Ministry of Water Resources, up 0.82% over the previous year; and 711 thousands working in local agencies, down 6.58%. The total salary of in-service staff members nationwide was 79.09 billion Yuan, and the annual average salary per person was 102 thousands Yuan.

Employees and Salaries

	2010	2011	2012	2013	2014	2015	2016	2017	2018	2019	2020
Number of in service staff /10^4 persons	106.6	102.5	103.4	100.5	97.1	94.7	92.5	90.4	87.9	82.7	77.8
Of which: staff of MWR and agencies under MWR /10^4 persons	7.4	7.5	7.4	7.0	6.7	6.6	6.4	6.4	6.6	6.6	6.7
Local agencies/10^4 persons	96.3	95.0	96.0	93.5	90.4	88.1	86.1	84.0	81.3	76.0	71.1
Salary of in-service staff /10^8 Yuan	297.9	351.4	389.1	415.3	451.4	529.4	640.5	739.1	802.7	787.6	790.9
Average salary /(Yuan/person)	28,816	34,283	37,692	41,453	46,569	55,870	69,377	83,534	91,307	95,385	102,000

Main Indicators of National Water Resources Development (2015–2020)

Indicators	unit	2015	2016	2017	2018	2019	2020
1. Irrigated area	10^3 ha	72,061	73,177	73,946	74,542	75,034	75,687
2. Farmland irrigated area	10^3 ha	65,873	67,141	67,816	68,272	68,679	69,161
Newly-increased in 2020	10^3 ha	1,798	1,561	1,070	828	780	870
3. Water-saving irrigated area	10^3 ha	31,060	32,847	34,319	36,135	37,059	37,796
Highly-efficient water-saving irrigated area	10^3 ha	17,923	19,405	20,551	21,903	22,640	23,190
4. Irrigation districts over 10,000 mu	Unit	7,773	7,806	7,839	7,881	7,884	7,713
Irrigation districts over 300,000 mu	Unit	456	458	458	461	460	454
Farmland irrigated areas in irrigation districts over 10,000 mu	10^3 ha	32,302	33,045	33,262	33,324	33,501	33,638
Farmland irrigated areas in irrigation districts over 300,000 mu	10^3 ha	17,686	17,765	17,840	17,799	17,994	17,822
5. Rural population accessible to safe drinking water	%	76	79	80	81	82	83
Centralized water supply system	%	82	84	85	86	87	88
6. Flooded or waterlogging area under control	10^3 ha	22,713	23,067	23,824	24,262	24,530	24,586

Continued

Indicators	unit	2015	2016	2017	2018	2019	2020
7. Controlled or improved eroded area	10^4 km^2	115.5	120.4	125.8	131.5	137.3	143.1
Newly-increased	10^4 km^2	5.4	5.6	5.9	6.4	6.7	6.4
8. Reservoirs	Unit	97,988	98,460	98,795	98,822	98,112	98,566
Large-sized	Unit	707	720	732	736	744	774
Medium-sized	Unit	3,844	3,890	3,934	3,954	3,978	4,098
Total storage capacity	10^8 m^3	8,581	8,967	9,035	8,953	8,983	9,306
Large-sized	10^8 m^3	6,812	7,166	7,210	7,117	7,150	7,410
Medium-sized	10^8 m^3	1,068	1,096	1,117	1,126	1,127	1,179
9. Total water supply capacity of water projects in a year	10^8 m^3	6,103	6,040	6,043	6,016	6,021	5,813
10. Length of dikes and embankments	10^4 km	29.1	29.9	30.6	31.2	32.0	32.8
Cultivated land under protection	10^3 ha	40,844	41,087	40,946	41,351	41,903	42,168
Population under protection	10^4 people	58,608	59,468	60,557	62,785	67,204	64,591
11. Total water gates	Unit	103,964	105,283	103,878	104,403	103,575	103,474
Large-sized	Unit	888	892	892	897	892	914
12. Total installed capacity by the end of the year	10^4 kW	31,937	33,153	34,168	35,226	35,564	36,972
Yearly power generation	10^8 kW·h	11,143	11,815	11,967	12,329	12,991	13,540
13. Installed capacity of rural hydropower by the end of the year	10^4 kW	7,583	7,791	7,927	8,044	8,144	8,134
Yearly power generation	10^8 kW·h	2,351	2,682	2,477	2,346	2,533	2,424
14. Completed investment of water projects	10^8 Yuan	5,452.2	6,099.6	7,132.4	6,602.6	6,711.7	8,181.7
Divided by different sources							
(1) Central government investment	10^8 Yuan	2,231.2	1,679.2	1,757.1	1,752.7	1,751.1	1,786.9
(2) Local government investment	10^8 Yuan	2,554.6	2,898.2	3,578.2	3,259.6	3,487.9	4,847.8
(3) Domestic loan	10^8 Yuan	338.6	879.6	925.8	752.5	636.3	614.0
(4) Foreign funds	10^8 Yuan	7.6	7.0	8.0	4.9	5.7	10.7
(5) Enterprises and private investment	10^8 Yuan	187.9	424.7	600.8	565.1	588.0	690.4

Continued

Indicators	unit	2015	2016	2017	2018	2019	2020
(6) Bonds	10^8 Yuan	0.4	3.8	26.5	41.6	10.0	87.2
(7) Other sources	10^8 Yuan	131.7	207.1	235.9	226.3	232.8	144.9
Divided by different purposes:							
(1) Flood control	10^8 Yuan	1,930.3	2,077.0	2,438.8	2,175.4	2,289.8	2,801.8
(2) Water resources	10^8 Yuan	2,708.3	2,585.2	2,704.9	2,550.0	2,448.3	3,076.7
(3) Soil and water conservation and ecological recovery	10^8 Yuan	192.9	403.7	682.6	741.4	913.4	1,220.9
(4) Hydropower	10^8 Yuan	152.1	166.6	145.8	121.0	106.7	92.4
(5) Capacity building	10^8 Yuan	29.2	56.9	31.5	47.0	63.4	85.2
(6) Early-stage work	10^8 Yuan	101.9	174.0	181.2	132.0	132.7	157.3
(7) Others	10^8 Yuan	337.5	636.2	947.5	835.8	757.4	747.3

Notes:

1. The data in this bulletin do not include those of Hong Kong, Macao and Taiwan.

2. Key indicators for water development and statistical data in 2012 and in 2013 is also integrated with the data of first national census for water.

3. Statistics of rural hydropower refer to the hydropower stations with an installed capacity of 50,000 kW or lower than 50,000 kW.

《2020年全国水利发展统计公报》编辑委员会

主　　　任：魏山忠
副 主 任：石春先
委　　　员：（以姓氏笔画为序）
　　　　万海斌　匡尚富　邢援越　巩劲标　朱　涛　朱闽丰
　　　　任骁军　刘六宴　刘宝军　孙　卫　李　烽　李兴学
　　　　李原园　吴　强　陈茂山　郑红星　姜成山　夏海霞
　　　　钱　峰　倪　莉　倪文进　徐永田　郭孟卓　郭索彦
　　　　曹纪文　曹淑敏　谢义彬

《2020年全国水利发展统计公报》主要编辑人员

主　　　编：石春先
副 主 编：谢义彬　吴　强
执 行 编 辑：汪习文　张光锦　乔根平
主要参编人员：（以姓氏笔画为序）
　　　　万玉倩　王　超　王位鑫　毕守海　曲　鹏　吕　烨
　　　　刘　品　刘宝勤　米双姣　许　静　孙宇飞　杜崇玲
　　　　李　益　李　森　李笑一　吴海兵　吴梦莹　邱立军
　　　　沈东亮　张　岚　张晓兰　张绪军　张雅文　房　蒙
　　　　郭　悦　戚　波　盛　晴
主要数据处理人员：（以姓氏笔画为序）
　　　　王小娜　王明军　王鹏悦　潘利业
英 文 翻 译：谷丽雅　侯小虎　张林若

◎ 主编单位
水利部规划计划司

◎ 协编单位
水利部发展研究中心

◎ 参编单位
水利部办公厅
水利部政策法规司
水利部财务司
水利部人事司
水利部水资源管理司
全国节约用水办公室
水利部水利工程建设司
水利部运行管理司
水利部河湖管理司
水利部水土保持司
水利部农村水利水电司
水利部水库移民司
水利部监督司
水利部水旱灾害防御司
水利部水文司
水利部三峡工程管理司
水利部南水北调工程管理司
水利部调水管理司
水利部国际合作与科技司
水利部综合事业局
水利部信息中心
水利部水利水电规划设计总院
中国水利水电科学研究院